Electronic Devices & Components

Electrical and Electronic Engineering

Engr Martin Nweke

Copyright © 2024 Engr Martin Nweke

All rights reserved.

ISBN: 9798884182738

DEDICATION

This book is dedicated to all Electrical and Electronic Engineering students worldwide!

CONTENTS

1 INTRODUCTION TO ELECTRONIC COMPONENTS

2 SEMICONDUCTORS

3 TRANSISTORS

4 THYRISTORS

5 POWER SOURCES

6 INTEGRATED CIRCUITS AND PROGRAMABLE DEVICES

7 BOOKS BY THE SAME AUTHOR

INTRODUCTION TO ELECTRONIC COMPONENTS

Electronic components are the basic units of every electronic circuit/ device. It is the systematic combination and arrangement of different components that makes up an electronic device. Electronics Engineers take advantage of the behavior of individual components and then arrange them to meet a specific need. The aim of such arrangement is usually to solve problems.

Electronic components are grouped into two major categories; the passive and active components.
An electrical component that dissipates, stores, and/or releases power instead of

producing it is known as a passive element. Coils (also known as inductors), capacitors, and resistances are examples of passive components.

An active element, or active component, is an electrical component that has the ability to energize the circuit. These components control supply and utilization of electricity in circuits. Basically, they control the flow of current through them.

Some of these components are current oriented /controlled while others are voltage oriented /controlled. As we move on, we will denote which components are current controlled and which are voltage controlled.

Energy sources (voltage or current sources), generators, alternators, semiconductor devices like transistors and photodiodes, etc. are a few typical examples of active circuit elements. The circuit's electric current flow is the responsibility of the active components.

We are going to have an in-depth analysis of electrical components under each category. Our main focus is on their working principles, characteristics, their behavior in circuits, and applications/ uses. We will as well treat simple calculations associated with these components.

Below is a list of electronic components/ elements grouped under active elements;
Semiconductors
Power Sources
Integrated Circuits
Programmable Devices, etc

SEMICONDUCTORS

Semiconductors as the name implies are partly conductors and partly insulators. Their conductivity lies between those of conductors and insulators. One unique feature of these components is that they can be made to behave like conductors and can influence how current flows through them. To understand semiconductors better, check SOLAR ELECTRICITY:Solar Energy, Semiconductors and Inverters *(see link at the 'books by same author' page).*

Examples of elements/ components under the category of semiconductors are

Diodes
Transistors
Thyristors

DIODES

An electrical component with two terminals that conducts electricity mostly in one direction is called a diode. On one end, it has little resistance, and on the other, high resistance.

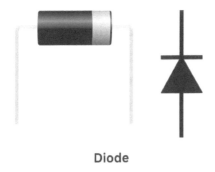

Diode

Working principle of diodes

A semiconductor is either P type or N type. When a P-type material is fused with an N-type, a junction is formed between the layers of these two materials. At every instance, electric current can flow from one layer to the other in only one direction into the connected part of an electric circuit. This is the basis of rectification; the most striking function of a diode. We all know that AC electricity is bi-directional but once it is made to flow through a diode, it loses one direction of flow due to this nature of diodes, hence current will flow in just that one direction into the connected parts of the circuit. These connected parts now function with the same type of electricity that comes from batteries. Once a diode permits current flow through in one direction, flow through the reverse direction is impossible. Passing electricity through a diode is called biasing. A situation when a diode and a voltage source is arranged in a position where flow of current through the diode is possible, it is known as forward biasing. The reverse arrangement where current will be unable to flow through a diode is known as reverse biasing. More details is given in the book about semiconductors.

Types of diodes

Below are some popular diode types and a brief insight to their working principles
Tunnel Diode
Backward Diode
Avalanche Diode
Transient Voltage Suppression (TVS) Diode
Gold Doped Diode
Constant Current Diode
Step Recovery Diode
Peltier Or Thermal Diode
Vacuum Diode
Varactor Diode
Gunn Diode
PIN Diode
Silicon Controlled Rectifier (SCR)
Shockley Diode
Point Contact Diode

P-N Junction Diode
This is one of the most basic semiconductor devices available, which has the electrical property of only allowing current to flow

through it in one direction. A diode, on the other hand, behaves differently from a resistor in response to applied voltage. Because of its exponential current-voltage (I-V) relationship, we are unable to adequately describe its behavior using a simple equation like Ohm's law.

P-N Junction

Small Signal Diode
Small signal diodes are tiny non-linear semiconductors that are frequently employed in digital logic, radio, and television circuits where high frequencies or tiny currents are involved. Compared to standard power diodes, compact signal diodes are smaller in size. The image and symbol of such diodes is the same as the regular power diodes, just that they are

smaller in size.

Schottky Diode

These are diodes formed by fusing a metal with an N-Type semiconductor to form a Metal-N junction instead of p-n junction. Metals like Aluminum and Platinum are mostly used for this purpose. They are mostly used in solar cells and another useful feature is their low capacitance due to negligible depletion region.

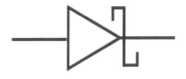

Schottky Diode

Super Barrier Diodes

They are Schottky diodes of the next generation. Applications requiring high power, low loss, and quick switching are intended for this diode. The little challenge with such diodes is their production cost.

Light Emitting Diode (LED)
These diodes emit visible light as soon as they get forward biased current pass through them). Holes from the N-region recombines with electrons from the N-region. When this happens, energy is released in form of photons. The energy needed for electrons to pass the semiconductor's band gap determines the color of the light.

They are mostly used for illumination and indication purposes.

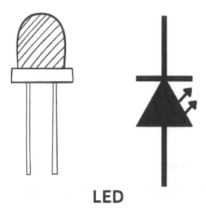
LED

Laser Diode

A laser diode is a semiconductor device similar to a light-emitting diode in which a diode pumped directly with electrical current can create lasing conditions at the diode's junction.

Laser diode

Zener Diode

This type of diode is specially designed to allow current flow in reverse polarity. Current flows backwards when the diode is biased with a voltage up to a certain set voltage known as zener voltage.

The most common application of zener diodes is in voltage regulation.

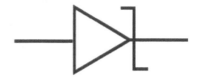

Zener Diode

Other diode types include
Rectifier Diode
Schottky Diode
Super Barrier Diodes
Tunnel Diode
Backward Diode
Avalanche Diode
Transient Voltage Suppression (TVS) Diode
Gold Doped Diode
Constant Current Diode
Step Recovery Diode
Peltier Or Thermal Diode
Vacuum Diode
Varactor Diode
Gunn Diode

PIN Diode
Silicon Controlled Rectifier (SCR)
Shockley Diode
Point Contact Diode

DIODE CALCULATIONS

Below is the diode equation

$i = I_S(e^{qv/kT} - 1)$

where i = current flowing through a diode

I_s = saturation current

q = charge on an electron = 1.602×10^{-19} C

k = Boltzmann constant = 1.38×10^{-23} joules/Kelvin

T = Temperature in Kelvin

TRANSISTORS

A semiconductor used to switch and amplify electrical impulses is called a transistor. It is a 3 terminal component with each terminal serving a unique purpose. The transistor terminals are;

Base terminal: Where the transistor is activated

Collector terminal: this is the positive terminal of the transistor

Emitter terminal: this is the negative terminal of the transistor

(img of a transistor)

PRINCIPLE OF OPERATION

Transistors serve two major purposes; signal amplification and switching. We are going to analyze the working principle of a transistor for these two major roles.

Transistors as amplifiers:

As an amplifier, a change in the base current results in a proportional change in the collector current (that handles the load); this is the basis of amplification. The common emitter configuration is often adopted while using a transistor as an amplifier. There is forward bias at the Base-emitter junction and reverse bias at the Collector-base junction. A little input signal is applied to the Base Emitter junction. This signal varies the voltage across the Base-emitter junction causing changes in the weight of the depletion region; with this, the base current changes.

Change in the base current is directly proportional to the change in the collector current leading to signal amplification.

$\Delta I_c \propto \Delta I_B$

The constant of this proportionality is the gain β

$\Delta I_C = \beta \Delta I_B$

The amplified signal appears across the

collector-emitter junction and the magnitude of this output is larger than that of the applied input.

In very simple terms when a transistor is energized, it is capable of producing a multiplied value of the magnitude of the signal applied at its input terminal. The good thing is that the applied input is usually very small and it yields much greater values at the output terminal. This makes transistors very useful in Electronic designs. They play major roles in sound amplification amidst other numerous uses.

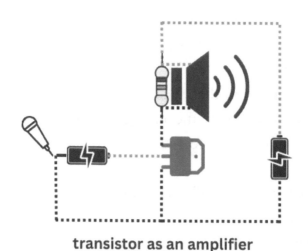

transistor as an amplifier

Transistor as a switch: considering the fact that the base of a transistor must be triggered with very little voltage before it can emit an output, loads connected to the output of a transistor can be switched by controlling what happens at the base of the transistor. Supply to the base can be fed from a light sensor for example. This means that whenever it senses light, the base of the transistor is supplied, and then whatever load is at the output terminal turns on. Transistors as switches are applicable to memory chips and computer processors.

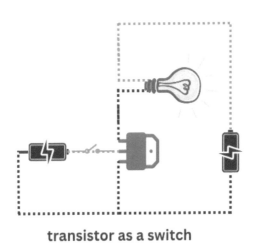

transistor as a switch

Types of transistors

Transistors are mainly grouped into BJTs and FETs. A BJT transistor is a setup of three semiconductor materials; two P-type materials and one N-type material, or two N-type materials and one P-type material. Hence we have PNP and NPN transistor structures. Transistors of this structure are known as BJTs

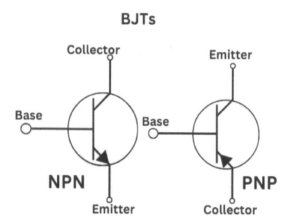

NPN

An NPN transistor is a particular kind of bipolar junction transistor (BJT) consisting of two N-type semiconductor layers sandwiched by a P-type layer. The arrangement of these semiconductor layers is where the word NPN originates.

The emitter, base, and collector are the three regions that are represented by the NPN transistor's three layers. Depending on the type of transistor, the area on one side that releases electrons or holes into the base is called the emitter. The center section, which is thin and mildly doped, serves as the basis. The area across from the base that gathers these charge carriers is known as the collector.

Since electrons make up the majority of charge carriers, their movement is essential to the functioning of an NPN transistor. A bigger current can flow from the collector to the emitter when a modest current (forward-biased) is delivered to the base-emitter junction (reverse-biased). Because of its ability to regulate a high output current with a low input current, the NPN transistor is a valuable tool for switching and amplification applications.

Whether studying semiconductor physics, constructing circuits, or troubleshooting devices, everybody involved in electronics has to understand the structure and behavior of NPN transistors.

Principle of Operation for NPN Transistors

An NPN transistor operates on the basis of controlling the base current to regulate the flow of current between the emitter and collector regions. Three terminals make up an NPN transistor: the base (B), the collector (C), and the emitter (E). N-type material is used for the emitter, P-type material is used for the base, and N-type material is used for the collector.

Electrons can move from the emitter to the base of the base-emitter junction when a tiny positive voltage is provided to the junction, causing it to become forward-biased. Few of these electrons recombine with holes in the base because the base area is narrow and mildly doped. Most of the electrons proceed to pass through the base and arrive at the reverse-biased collector.

Electrons are drawn toward the collector and repelled away from the base by the electric field produced by the reverse-biased base-collector junction. The lesser base current regulates the big current that flows from the collector to the emitter as a result. The transistor's current gain (β) is defined as the ratio of the collector current to the base current.

Depending on the biasing circumstances, the NPN device functions in three distinct regions:

The base-emitter junction is forward-biased and the base-collector junction is reverse-biased in the active region. The output current of the transistor is proportionate to the input current, acting as an amplifier.

Cutoff region: Both the base-collector and base-emitter junctions are reverse-biased in this area. There is no current flowing between the collector and emitter of the transistor when it is in the "off" state.

The base-emitter and base-collector junctions are both forward-biased in the saturation zone. When the transistor is in its "on" state, the collector and emitter are receiving the maximum amount of current.

PNP

Another variety of bipolar junction transistor (BJT) that is almost the NPN transistor's mirror image is the PNP transistor. It is made up of an N-type semiconductor layer sandwiched between two P-type semiconductor layers. The arrangement of these semiconductor layers is where the word PNP originates.

The emitter, base, and collector are the three regions that are represented by the PNP transistor's three layers. The area on one side that releases holes into the base is known as the emitter. The center section, which is thin and mildly doped, serves as the basis. The area across from the base that gathers these charge carriers is known as the collector.

With the roles of the electrons and holes interswitched, a PNP transistor operates similarly to an NPN transistor. A greater current can flow from the emitter to the collector (forward-biased) when a smaller current is applied to the reverse-biased base-emitter junction. The PNP transistor is a helpful tool for switching and amplification because of its ability to manage a large output current with a modest input current.

PNP transistor structure and operation are essential for any electronics enthusiasts, be it circuit design, device troubleshooting, or semiconductor physics research.

Principle of Operation for PNP Transistors
Similar to its NPN counterpart, a PNP transistor operates on the idea of controlling the flow of current between the emitter and

collector regions through adjustments to the base current. In contrast to the NPN transistor, which uses electrons as the majority charge carrier, the PNP transistor uses holes as the majority charge carrier.

A little negative voltage given to the base-emitter junction of a PNP transistor causes the junction to become forward-biased. This permits holes to go from the P-type material emitter to the N-type material base. Only a small portion of these holes recombine with electrons in the base because the base region is narrow and mildly doped. Most of the holes continue to flow through the base and into the reverse-biased collector.

The holes are drawn toward the collector and repelled away from the base by the electric field produced by the reverse-biased base-collector junction. The lesser base current regulates the huge current that flows from the emitter to the collector as a result. The transistor's current gain (β) is defined as the ratio of the collector current to the base current.

Depending on the biasing circumstances, the PNP transistor functions in three distinct regions:

The base-emitter junction is forward-biased and the base-collector junction is reverse-biased in the active region. The output current of the transistor is proportionate to the input current, acting as an amplifier.

Cutoff region: Both the base-collector and base-emitter junctions are reverse-biased in this area. There is no current flowing between the emitter and collector of the transistor when it is in the "off" state.

The base-emitter and base-collector junctions are both forward-biased in the saturation zone. When the transistor is in its "on" state, the emitter and collector are receiving the maximum amount of current.

A common form of transistor used for weak-signal amplification (e.g., amplifying wireless communications) is the field-effect transistor (FET). It has the ability to amplify analog and digital signals. It can also act as an oscillator or switch DC.

The operation of a field-effect transistor (FET)

Current travels along the channel, a semiconductor route, in the FET. An electrode known as the source is located at one end of the channel. The drain electrode is located at the opposite end of the channel. The channel's effective electrical diameter can be changed by applying a voltage to the gate, a control electrode, even though the channel's physical diameter is fixed. At any given point in time, the electrical diameter of the channel determines the conductivity of the FET. The current flowing from the source to the drain can vary greatly in response to even a tiny change in the gate voltage. The FET amplifies impulses in this way.

Field-effect transistors can be divided into two main categories. These are referred to as metal-oxide-semiconductor FETs (MOSFETs) and junction FETs (JFETs).

In the case of a Junction FET, it has a channel that consists of a P-type semiconductor (for a P-channel FET) or an N-type semiconductor (for an N-CHANNEL FET), then the gate is made of an opposite semiconductor material to the semiconductor material of the channel. Electric charges are mostly transported in P-type material as electron deficits known as holes. Electrons are the main charge carriers in N-type material. The junction in a JFET is the line dividing the channel from the gate. This P-N junction is normally reverse-biased, meaning that a DC voltage is supplied to it, preventing any current from passing between the gate and the channel.

The channel of a MOSFET may consist of a P-type or N-type semiconductor. An oxidized metal piece serves as the gate electrode. The oxide layer separates the gate from the channel electrically. Because of this, the MOSFET was once referred to as the insulated-gate FET (IGFET), albeit this nomenclature is no longer frequently used. There is practically never any current flowing between the gate and the channel at any point throughout the signal cycle because the oxide layer serves as a dielectric. The MOSFET has a very high input impedance as a result. The MOSFET is vulnerable to electrostatic charge damage due to the incredibly thin oxide layer. MOS devices must be handled and transported with extra attention.

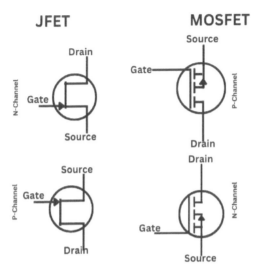

Applications of Jfet

Voltmeters, oscilloscopes, and other measuring instruments are examples of electrical equipment that commonly use FETs as input amplifiers. Radiofrequency amplifiers for FM appliances are also made using FET. TV and FM receivers operate as mixers thanks to FET technology.

Some useful transistor formulas
$IE = IC + IB$

This means that the Emitter current equals the sum of the Collector and Base current.

$I_C / I_E = \alpha$

This formula explains that the common base AC current gain (α) of a transistor is the ratio of the variation in collector current to the variation in the emitter current.

The value of α is ideally =1.

$I_C / I_B = \beta$

This formula explains that the common Emitter AC current gain (β) of a transistor is the ratio of the variation in collector current to the variation in the base current.

The value of β is ideally far greater than 1.

$\alpha = \beta / 1+ \beta$ (in terms of α)
$\beta = \alpha / 1-\alpha$ (in terms of β)

The above formulas describe the relationship between α and β

CONSIDERING COMMON VOLTAGES V_{CE}, V_{BE}, V_{CB}, V_{CC}

Observe the diagram below

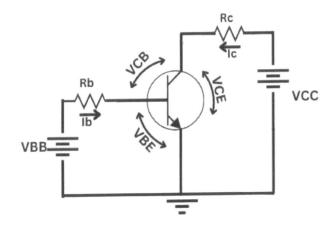

V_{BE} is the voltage across the base and the Emitter and has a constant value of 0.7v

V_{CE} is the voltage across the collector and emitter and is given by

$V_{CE} = V_{CC} - I_C \times R_C$ (V_{CC} is collector supply voltage)

V_{CB} is the voltage across the base and the collector, given by

$V_{CB} = V_{CE} - V_{BB}$; Where V_{BB} is the supply voltage to the Base of the transistor.

THYRISTORS

A thyristor is a semiconductor device with four layers that alternates between P-type and N-type materials (PNPN). Three electrodes are typically included in a thyristor: an anode, a cathode, and a gate, sometimes referred to as a control electrode.

It is a solid-state semiconductor device used for high-power applications. Unlike transistors that switch milliamps of DC current, Thyristors are capable of switching 10-15 amps of alternating current

The working principle of thyristors are similar to that of transistors but there are minor differences. When a low trigger current flows from the Gate pin of a thyristor, a current flows from the anode pin to the cathode pin. This means that with a low trigger current from the gate pin, a high current between the anode and cathode is controlled. While the current controlled here is generally small in the case of transistors, current of magnitude ranging from Ma to A values are controlled in thyristors.

The anode and cathode terminals are usually connected in series with the load to which power is to be controlled. For thyristor circuits to have the ability of delivering large currents, they must be able to withstand large externally applied voltage.

All thyristor types are controllable in switching from a forward-blocking state into a forward-conduction state. Most thyristors have a characteristic that after switching from a forward-blocking state into a forward-conduction state, the gate signal can be removed and the thyristor will remain in its forward-conduction mode. This property is known as LATCHING.

This is the major differentiating factor between thyristors and other power electronic devices. In some cases, there are thyristors that can be controllable in switching from forward-conduction back to forward-blocking. The manufacturers design determine the controllability and application of every given thyristor.

Just like diodes, a thyristor is also a unidirectional device and can be made to operate as either an open circuit switch or as a rectifying diode depending on how the thyristor gate is triggered. They can only operate in switching mode and cannot be used for amplification.

Summarily, a thyristor is a three terminal component which can be switched ON and OFF at extremely fast rate, or it can be switched ON for variable duration during half cycles to deliver a specific amount of power to a load.

Just like diodes, thyristors can be forward and reverse biased.

Thyristor calculations

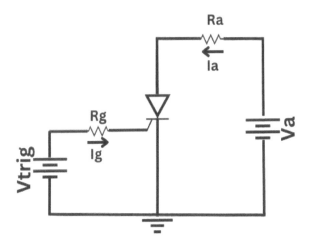

V_{trig} = Gate voltage

R_G = Gate resistance

V_{Gk} = Voltage across gate cathode junction

I_G = Gate current

V_A = Anode voltage

I_A = Anode current

R_A = Anode Resistance

V_{AK} = Voltage across anode and cathode junction

$V_{trig} = I_G R_G + V_{Gk}$

$I_G = (V_{trig} - V_{Gk})/ R_G$

$V_A = I_A R_A + V_{Ak}$

$I_A = (V_A - V_{Ak})/ R_A$

POWER SOURCES

Every electrical appliance or gadget we use has a power rating indicated on it. It implies that it uses that particular rated power for useful energy conversions in the circuit. For example, a cellphone's display unit (which produces light), speakers (which produce audio), processors (which do logical operations), and other components are all powered by the battery. Some generate heat energy by converting electrical energy from electricity sources.

Power is the rate at which work is done or the rate at which electric energy is transferred in an electric circuit. This work is carried out by electric current which is propelled by an electric force (voltage). The combined effect of both voltage and current in a circuit is the power rating.

In electrical circuits, we mostly see voltage (DC or AC) and current source symbols. The power consumed in these circuits is drawn from these sources, mostly batteries or solar cells (for DC) and generators, or public grid supplies (for AC). Below are some of these sources and their symbols.

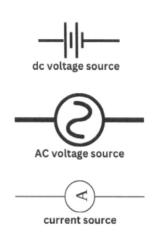

dc voltage source

AC voltage source

current source

INTEGRATED CIRCUITS AND PROGRAMABLE DEVICES

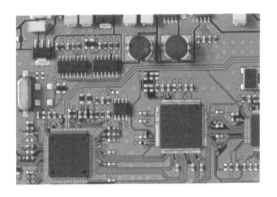

A microchip, also known as an integrated circuit (IC), is a tiny electronic circuit array created by fabricating different electrical and electronic components (such as transistors, capacitors, and resistors) on a silicon wafer. These chips can carry out operations akin to those of large discrete electronic circuits composed of discrete electronic components.

It is called an integrated circuit, integrated chip, or microchip because all these arrays of components, tiny circuits, and semiconductor wafer material basis are merged to form a single chip.

Individual or discrete electronic components of varying sizes are utilized to create electronic circuits; as a result, the cost and size of these discrete circuits grow as the number of components employed in the circuit increases. In order to combat this negative aspect, integrated circuit technology was developed. In the 1950s, Jack Kilby of Texas Instruments created the first integrated circuit or IC. Robert Noyce of Fairchild Semiconductor then solved some of the integrated circuit's practical issues.

Every electrical and electronic gadget that we use on a daily basis, including computers, laptops, refrigerators, televisions, and mobile phones, is made with some kind of circuit, no matter how complicated. Multiple electrical and electronic components, such as resistors, capacitors, inductors, diodes, transistors, and so forth, are connected to one another by connecting wires or conducting wires to allow electric current to flow through the circuit's multiple components. Circuits can be categorized into various types according to various factors. For example, circuits can be classified as series or parallel based on connections; as integrated or discrete circuits based on size and manufacturing process; and as analog or digital circuits based on the type of signal used in the circuit.

Various Integrated Circuit Types

There are several varieties of integrated circuits (ICs), and they are categorized using a number of distinct factors.

The IC is categorized as an analog integrated circuit, a digital integrated circuit, or a mixed integrated circuit depending on the intended use.

Digital integrated circuits

Digital ICs are integrated circuits that function at a limited number of defined levels rather than at all possible signal amplitude levels. They are created by combining several digital logic gates, multiplexers, flip flops, and other electrical circuit components. Binary or digital input data, such as 0 (low or false, or logic 0) and 1 (high or true, or logic 1), can be sent into these logic gates.

Linear ICs

An analog integrated circuit is referred to be linear IC if there is a linear relationship between its voltage and current. The 7.41 IC, an 8-pin DIP (Dual In-line Package) op-amp, is the best illustration of this type of linear integrated circuit.

Analog ICs

Analog integrated circuits (ICs) are those that function across a continuous signal range. They are separated into two categories: radio frequency integrated circuits (RF ICs) and linear integrated circuits (Linear ICs). In fact, across a wide range of continuous analog signals, the relationship between the voltage and current may not always be linear.

An operational amplifier, or op-amp for short, is a commonly used analog integrated circuit that functions similarly to a differential amplifier but with a much higher voltage gain. Compared to digital ICs, it has a very small number of transistors. Computerized simulation techniques are used to build analog application-specific integrated circuits or analog ASICs.

Mixed ICs

Mixed integrated circuits (ICs) are created by combining digital and analog integrated circuits on a single chip. These integrated circuits perform the roles of clock/timing, analog to digital (A/D and D/A), and digital to analog converters. The circuit pictured above is an illustration of a mixed integrated circuit and shows a self-healing radar receiver operating at 8 to 18 GHz.

RF Integrated Circuits

Radiofrequency ICs are analog integrated circuits (ICs) that have a non-linear relationship between their voltage and current. An alternative name for this type of IC is a radio frequency integrated circuit.

The development of integration technology has allowed for the integration of digital, multiple analog, and RF operations on a single chip, leading to the creation of mixed-signal systems on a chip.

The following are examples of general integrated circuit (IC) types:

Comparators: The comparator integrated circuits (ICs) serve as comparators to compare the inputs and subsequently provide an output based on the comparison of the ICs.

Logic circuits

These integrated circuits are made with logic gates, which have binary input and output (0 or 1). They serve mostly as decision-makers. All of the logic gates connected inside the integrated circuit (IC) provide an output based on the circuit connected inside the IC, which is designed to be used for carrying out a certain task. This output is determined by the logic or truth table of the logic gates.

ICs for voltage regulators

Notwithstanding variations in the DC input, this type of integrated circuit produces a steady DC output. Examples are LM309, uA723, LM105, and 78XX ICs.

Integrated Circuit CMOS

Compared to FETs, CMOS integrated circuits are used in a wide range of applications due to their lower threshold voltage and low power consumption. P-MOS and N-MOS transistors that are co-fabricated on the same chip are included in CMOS integrated circuits. This integrated circuit's design is a Polysilicon gate, which helps to lower the threshold voltage of the apparatus and enable low-voltage operation.

IC timers

Timers are specialized integrated circuits used in intended applications for counting and timekeeping purposes.

Speaker Amplifiers

Among the many different kinds of integrated circuits (ICs) utilized for audio amplification are audio amplifiers. These are typically utilized in television circuits, audio speakers, and other devices. The circuit for the low-voltage audio amplifier IC is shown above.

Functioning Amplifiers

Like audio amplifiers, which are used for audio amplification, operational amplifiers are widely utilized integrated circuits. These integrated circuits (ICs) function similarly to transistor amplifier circuits and are used for amplification.

Integrated Circuit Types according to Classes

Depending on the methods utilized in their manufacture, integrated circuits are divided into three types.

Single-unit integrated circuits
ICs with thick and thin films
ICs that are hybrid or multichip

Single-unit integrated circuits

These integrated circuits allow for the formation of connections between discrete, passive, and active components on a silicon chip. These ICs are currently the most often utilized since they are more dependable and less expensive. Commercially produced integrated circuits (ICs) are utilized in computer circuits, AM receivers, amplifiers, and voltage regulators. Nevertheless, the monolithic IC components have a lower power rating and inadequate insulation.

Thick & Thin ICs

Passive components such as resistors and capacitors are employed in these kinds of integrated circuits, but the transistors and diodes are connected as independent components in order to create a circuit. These integrated circuits (ICs) are essentially combinations of separate and integrated components, and they share similar properties and appearances other than the method of film deposition. The thin ICs film deposition can be chosen based on ICS.

These integrated circuits are created by depositing sheets of conducting material either on a ceramic stand or on the surface of glass. It is possible to manufacture passive electronic components by varying the resistivity of the materials by altering the film thickness.

The appropriate circuit model is created on a ceramic substrate in this kind of integrated circuit using the silk printing technique. Printed thin-film integrated circuits are another name for these chips.

Dual-Chip or Hybrid ICs

Multi refers to more than one linked chip, as the name would imply. These integrated circuits (ICs) are considered passive components, while diffused capacitors or resistors on a single chip are considered active components, such as diodes or diffused transistors. Metalized prototypes are a possible means of connecting these parts. Multi-chip integrated circuits are widely used in high-power amplifier applications ranging from 5W to 50W. The performance of hybrid integrated circuits is better than that of monolithic ICs.

Kinds of IC Packing

Two types of IC packages are distinguished:

Through-Hole Mount and Surface Mount Packaging.

Whole-Hole Mounting Sets

These can be designed so that the lead pins are welded through one side of the board and smoldered through the other. These packages are larger in size than others of their kind. These are mostly used in electrical devices to balance cost and board space constraints. Dual inline packages are the most often used through-hole mount packages, making them the finest example. There are two varieties of these packages: ceramic and plastic.

The 28 pins of the ATmega328 are arranged on a black plastic rectangular-shaped board and are positioned parallel to one another by expanding vertically. A 0.1-inch gap is maintained between each pin. Furthermore, the variation in the number of pins in different packages causes the package to alter in size. It is possible to position these pins so that they may be controlled to the center of a breadboard, preventing short circuits.

The many through-hole mount integrated circuit packages include T7-TO220, TO2205, TO220, TO99, TO92, TO18, and TO03.

Packaging for Surface Mount Systems

This type of packaging primarily adheres to mounting technology, unless the components are directly located on the PCB. His manufacturing techniques will speed up the process, but because the parts are small and closely spaced, there is a greater possibility of errors. In this type of packaging, plastic or ceramic molding is adopted. The tiny outline L-leaded package and BGA (Ball Grid Array) are two types of surface-mount packaging that use plastic molds.

SOT23, SOT223, TO252, TO263, DDPAK, SOP, TSOP, TQFP, QFN, and BGA are the many surface mount integrated circuit packages.

PROGRAMMABLE DEVICES

Programmable devices are electronic components that may be adjusted and modified to carry out particular functions according to the user's needs. A notable instance is the Programmable Logic Controller (PLC), extensively utilized in industrial automation. Programmable Logic Controllers (PLCs) oversee and regulate machinery and operations in manufacturing settings. These devices use input and output modules to connect with sensors and actuators, performing logic functions to automate processes effectively.

Field Programmable Gate Arrays (FPGAs) and Application-Specific Integrated Circuits (ASICs) are other types of programmable devices. FPGAs provide reconfigurable circuitry that can adjust to evolving needs, making them ideal for prototype and quick development. ASICs are specialized integrated circuits tailored for specific tasks, providing superior performance and efficiency but without the capacity to be reprogrammed.

Programmable gadgets go beyond conventional electronic hardware. Computers are programmable machines that can execute sequences of instructions and are the basic components of modern computing. Through software programming, computers execute varied jobs ranging from simple computations to complicated simulations and data processing.

Programmable devices are essential for many technical developments. They are key components in consumer electronics, telecommunications systems, and automotive applications. From microcontrollers incorporated in household appliances to sophisticated control systems in aerospace engineering, programmable electronics allow innovation and automation across industries.

It encompasses a broad variety of electronic components and systems meant to be adaptable and customizable. These gadgets offer automation, customization, and invention across many industries, driving technological development and efficiency.

Programmable devices, such as Programmable Logic Devices (PLDs), play a significant part in modern electronics. PLDs consist of arrays of logic gates, often AND and OR gates, which can be customized to execute certain operations. These devices enable flexibility and adaptability in circuit design, allowing engineers to execute complicated logic operations without the requirement for custom-designed integrated circuits.

Field Programmable Gate Arrays (FPGAs) are another form of programmable device widely employed in numerous applications. FPGAs consist of an array of adjustable logic blocks interconnected via programmable routing resources. They offer considerable flexibility and performance, making them useful for applications such as signal processing, networking, and embedded systems.

Complex Programmable Logic Devices (CPLDs) are programmable logic devices that offer a smaller scale of integration compared to FPGAs but with quicker operation speeds. CPLDs are widely utilized in situations where real-time processing and low power consumption are required. They find uses in consumer electronics, automobile electronics, and industrial control systems.

BOOKS BY THE SAME AUTHOR

A-Z of Single phase generators and electric motors: Practical Electrical job guide

https://www.amazon.com/dp/B0BH31J2VX

SOLAR ELECTRICITY: Solar Energy, Semiconductors and Inverters

https://www.amazon.com/dp/B0C2RJT8VX

Electrical Engineering Career: Becoming an Electrical Engineer

https://www.amazon.com/dp/B0CCZWMSHX

Electromagnetic Theory for complete novices

https://www.amazon.com/dp/B0BTRW3BJ3

CCTV CAMERA INSTALLATION: Analog and IP Camera Surveillance Systems

https://www.amazon.com/dp/B0BSJPZVBC

Advanced CCTV Surveillance Systems: modern surveillance systems

https://www.amazon.com/dp/B0CRQL8BFN

CCTV Surveillance Systems: Faults, and Solution measures

https://www.amazon.com/dp/B0CRPRHBRM

Electronic Devices & Components

Made in United States
Troutdale, OR
10/07/2024

23478939R00040